BEYOND DELIVERY

HOW TECHNICAL PROGRAM MANAGERS LEAD CHANGE AND GROW WITH IMPACT

JOHNATHAN STEPHEN SEXTON

"Leadership is clarity in action — influence that turns confusion into momentum."

Beyond Delivery: How TPMs Lead Change and Grow with Impact

By Johnathan Stephen Sexton

Dedication

This book is dedicated to the many partners I've had the privilege of working with throughout my career. From supply chain to technology, from store operations to corporate strategy, Health and Wellness to Enterprise Business Services, every team and every collaboration has taught me that you never arrive; learning is constant.

A very heartfelt *"Thank You"* to the many leaders I've had the privilege of learning from over the years.

Thank you for teaching me, challenging me, and reminding me that leadership is never done alone. Your influence and partnership are part of every page in this book.

Copyright & Disclaimer

Beyond Delivery: How Technical Program Managers Lead Change and Grow with Impact

Copyright © 2025 by Johnathan Stephen Sexton

Disclaimer

This book is designed to provide general information and leadership insights. It is not intended as professional, legal, business, financial, medical, or fitness advice. The author and publisher disclaim liability for any losses, damages, or injuries that may result from the application of ideas, practices, or information contained in this book. Readers should seek professional guidance appropriate to their individual circumstances.

Affiliation Notice

The views and opinions expressed in this book are solely those of the author. They do not represent the views of Walmart Inc., Maxwell Leadership, NASM, or any other organization with which the author is or has been affiliated. Mention of these organizations does not imply endorsement.

Trademarks

All trademarks, registered trademarks, product names, and company names or logos mentioned in this book are the property of their respective owners. Use of these names does not imply any affiliation with or endorsement by them.

Scripture Notice

ISBN (Paperback): 979-8-218-79328-9
ISBN (eBook): [to be assigned]

Printed in the United States of America.
First Edition, 2025

OneChristianMan.com

Introduction — Why TPM Leadership Matters Now

In most organizations, value is unlocked at the seams, where strategy meets execution, product meets engineering, and business constraints meet technical reality. Those seams are messy: priorities shift, dependencies collide, and no single person "owns" the whole picture. This is where Technical Program Managers create disproportionate impact.

We don't lead because we hold formal authority over people or budgets; we lead because we orchestrate clarity, alignment, and momentum across complex systems.

What Is a Technical Program Manager (TPM)?

A Technical Program Manager (TPM) is a cross-functional leader responsible for guiding complex, high-stakes initiatives across engineering, product, operations, and business functions. Unlike traditional managers, TPMs rarely have direct authority over teams. Instead, they lead by creating clarity and alignment.

Over the last decade, the TPM role has shifted from "project traffic controller" to "force-multiplier leader." Agile at scale, platform thinking, AI-accelerated workflows, and continuous delivery have raised the bar. Organizations need leaders who can translate their vision into coordinated action, remove friction at the portfolio level, and help teams make better decisions more quickly. That's the job. And it matters now more than ever.

The Leadership Gap We're Here to Fill

Most companies don't struggle because they lack ideas. They struggle because they lack alignment and consistent follow-through. The result is a set of common gaps that quietly drain momentum.

One of the biggest challenges is **decision latency**. Teams wait for direction, while risk grows and costs accumulate in the background. Another common issue is **diffuse ownership**. When everyone is responsible, no one feels truly accountable, and progress stalls. There is also **local optimization**. Functions chase their own metrics, but in doing so, they sacrifice end-to-end outcomes.

Finally, companies accumulate what has been called **narrative debt**. Leaders lose sight of the system itself, so they fall back on "status theater" — polished updates that look good in the moment but fail to reveal the truth.

These gaps are rarely intentional, but left unaddressed, they prevent organizations from turning good ideas into lasting impact. Effective TPMs address gaps by clarifying decisions, ownership, constraints, and system context so leaders can act confidently.

From Delivery to Impact

Shipping features isn't enough; impact matters more than deadlines.

Strong TPMs:

1. Define the problem and key tradeoffs early.
2. Set up decision-making routines and highlight risks.
3. Use essential metrics to track reality.
4. Communicate goals clearly across teams.
5. Streamline process, scope, dependencies, and ownership for faster results.

What This Book Will Help You Do

By the end, you'll have what's necessary to:

- Influence and build trust quickly, even without authority.
- Shift from reactive to proactive systems leadership.
- Use AI and data effectively to boost results.
- Expand your impact beyond individual projects.
- Build a leadership brand to advance your career.

The TPM Leadership Flywheel

As the flywheel spins, the system gets faster and simpler; your credibility compounds; your scope and influence grow.

This book presents a practical flywheel for any program:

- **Clarity:** Define goals, limits, and decision-makers.
- **Cadence:** Set up simple routines to maintain progress.
- **Visibility:** Make the actual status clear and eliminate unnecessary reporting.
- **Enablement:** Actively remove barriers through targeted experiments.
- **Narrative:** Turn updates into actionable, executive-level stories.

At its core, this book is about moving beyond the mechanics of delivery to embrace the true calling of a TPM: leading change and enabling growth through clarity, influence, and resilience. The frameworks and stories ahead will help you transform ambiguity into opportunity, decisions into momentum, and visibility into trust. Whether you're stepping into your first TPM role or scaling impact at the enterprise level, the path forward is not about managing tasks —

it's about multiplying outcomes. That is the work of leadership, and it begins now.

Part 1: Leading Without Authority

"Outstanding leaders go out of their way to boost the self-esteem of their personnel. If people believe in themselves, it's amazing what they can accomplish."
— Sam Walton

In nearly every program, the Technical Program Manager sits at the center of decisions without being the official "owner" of them. We don't sign the budget, we don't own the product roadmap, and we don't push code. Yet we're accountable for helping all of it come together.

This is the paradox — TPMs are responsible for outcomes, but we rarely hold formal authority. Success depends on our ability to lead through influence, clarity, and consistency.

Why this matters now:

- Tech initiatives are increasingly **cross-functional** — requiring marketing, product, design, engineering, compliance, and operations to coordinate at speed.
- Formal hierarchies can't keep up with the **pace of change**, but decisions still need to be made quickly.
- People resist being "managed" but respond to those who **make their work easier and more successful.**

Leading without authority is not about charm or persuasion tricks. It's about building **trust** and **credibility** through repeated delivery of value: clarifying goals, removing friction, and framing the story so teams and leaders can align.

In this section of the book, you'll learn:

1. The Invisible Leader — why TPMs often feel overlooked, and how to turn invisibility into strategic presence.
2. Influence Is the Currency — the real mechanics of trust and credibility across teams.

3. Driving Decisions in the Gray Zone — how to navigate ambiguity and guide outcomes when nobody "owns" the answer.

This is the **core craft of TPM leadership**. Before we can scale, grow, or think about futureproofing, we must master this foundational ability: to lead change when authority isn't granted.

Chapter 1: The Invisible Leader

"A leader is one who knows the way, goes the way, and shows the way."
— *John C. Maxwell*

The Paradox of TPM Leadership

If you've been a TPM for more than a few months, you've likely felt invisible. You don't write code, yet engineering depends on you. You don't own the roadmap, yet product counts on you. You don't sign the budget, but finance expects you to know the impact of delays. You're everywhere and nowhere all at once.

The paradox of TPM leadership is this: you're accountable for results without being the visible "owner" of them. That invisibility can feel frustrating. But it's also your advantage.

Why Invisibility Can Be an Asset

Because TPMs aren't bound to a single team or discipline, we see the system in ways others cannot. Product managers are focused on customers, engineers on architecture, designers on usability, finance on costs. As a TPM, you stitch these perspectives together into one coherent story.

Your "invisibility" allows you to:

- **Spot dependencies** nobody else is watching.
- **Translate across disciplines** without bias toward one function.
- **Surface risks early** without being seen as defending a single silo.

- **Create alignment** because you're not competing for the same credit.

Invisibility is the camouflage that allows TPMs to move through organizations, identifying and fixing the seams.

Field Story: Lessons from the East Data Center

In 2004, I wasn't yet a TPM. I was a programmer, writing code to support Walmart's dynamic distribution system as it was being integrated into cross-dock replenishment. I wasn't a project lead — I was part of a focused group of developers tasked with building something that would reshape replenishment for the company.

The project was so important that leadership physically moved our team into a separate room at the East Data Center. They wanted us out of the day-to-day noise, entirely dedicated to making this system work. The unfortunate part is that this space had no windows, and the restroom door was literally next to our work area. Every time the toilet flushed, everyone would hear it.

At the time, I didn't realize it, but this was my first exposure to the seams where leadership really happens. Even though I was "just coding," I could see how every decision connected — replenishment forecasts, trucking schedules, store inventory levels. We were deep in the details, but the impact was massive. The system we were coding would later increase store in-stock by 5% chainwide.

Looking back now, I see how much **invisible leadership** was already at work. TPMs (though we didn't call them that yet) were pulling together logistics, replenishment, store operations, and IT. They did write some code, but they also framed tradeoffs, set cadences, and made sure our work aligned with business impact.

That season taught me a valuable truth: sometimes the most important leaders aren't in the spotlight. They're in the seams, stitching the organization together so the builders can focus and deliver.

Shifting from Coordinator to Leader

Too many TPMs settle into the role of "meeting note-taker" or "status coordinator." That's where invisibility becomes dangerous. You're present, but not impactful.

Leadership begins when invisibility turns into **strategic presence**:

- Instead of just running the meeting, frame the decision that needs to be made.
- Instead of distributing status, synthesize insights that shift priorities.
- Instead of tracking tasks, highlight risks and tradeoffs that matter.

Leadership is measured not by how many meetings you schedule, but by the clarity you create.

Maxwell Application: The Law of the Mirror

John Maxwell teaches in *The 15 Invaluable Laws of Growth* that *"You must see value in yourself to add value to others."* This law is crucial for the invisible leader. When your role isn't celebrated on an org chart, you can easily underestimate your own contribution.

But invisibility doesn't equal insignificance. When you believe in the value of your leadership — your ability to create clarity, align teams,

13

and guide decisions — you'll project confidence even when others overlook your role.

Walton Application: Belief as a Catalyst

Sam Walton echoed this truth when he said: "Outstanding leaders go out of their way to boost the self-esteem of their personnel. If people believe in themselves, it's amazing what they can accomplish."

But Walton's principle comes with a challenge: before you can instill belief in others, you must first believe in the value you bring. Leaders who doubt themselves project hesitation; leaders who stand firm in their own worth create the confidence that multiplies belief in their teams.

For TPMs, that means showing up with the quiet conviction that your clarity, influence, and presence matter. When you carry that confidence, you unlock the same belief in those you lead — and suddenly, they accomplish more than they thought possible.

Building Credibility When Nobody Sees You

Without formal authority, credibility is your currency. People follow TPMs they trust. Three levers matter most:

1. **Competence**: You understand the technical architecture, delivery processes, and business goals well enough to connect the dots.
2. **Consistency**: You show up the same way every time: prepared, reliable, calm under pressure.
3. **Contribution**: You add value in every interaction, even small ones: the correct link, the clarified scope, the distilled risk.

When you consistently demonstrate these, you go from invisible to indispensable.

Reflection Exercise: From Invisible to Indispensable

Take 15 minutes this week to complete this reflection:

1. Write down three ways your work currently feels "invisible."
2. For each, ask: *How could I turn this into a moment of presence?* (e.g., reframing a meeting, simplifying a report, clarifying a decision).
3. Identify one credibility lever (competence, consistency, contribution) you can strengthen this week.

The key is not to fight invisibility — but to transform it into trust and presence.

Key Takeaways

- Invisibility isn't a weakness; it's an advantage when leveraged.
- TPM leadership starts when coordination evolves into clarity creation.
- Credibility (competence, consistency, contribution) turns invisibility into influence.
- Your impact isn't measured in tasks completed, but in outcomes enabled.

Chapter 2: Influence Is the Currency

"Leadership is influence—nothing more, nothing less."
— John C. Maxwell

Leadership Without a Badge

In most organizations, authority comes with a title, such as Senior Director, VP, or CTO. But Technical Program Managers rarely wield positional power. Your authority isn't printed on an org chart; it's earned. The lever that TPMs pull most effectively isn't authority — it's influence.

Influence is the ability to move decisions forward, align stakeholders, and shape outcomes when you can't command or control. It's the actual currency of TPM leadership.

What Influence Really Looks Like

Influence often gets mistaken for charisma or politics. In reality, it's built through repeated deposits of trust into your credibility account. When people trust you, they let you guide decisions.

Influence as a TPM shows up in small but powerful ways:

- A senior engineer rethinks their approach because you highlight a dependency they missed.
- A product leader changes a priority after you frame the tradeoffs in business terms.
- An executive makes a call faster because your summary gives them confidence.

Each of these moments is earned, not granted.

Field Story: Dynamic Distribution of Perishables

By 2005, I had transitioned from pure programming into broader program roles, working on Walmart's **dynamic distribution system** as it was rolled out into **perishable replenishment**. This wasn't just another IT upgrade — it was a project that directly affected the freshness of produce and the trust of millions of customers.

The tension was constant.

- **Logistics** wanted efficiency through scheduled truck routes and optimized loads.
- **Merchants** wanted to drive aggressive sales, even if it meant uneven distribution.
- **Store operations** wanted to guarantee customers could find fresh produce on every visit.

No single team "owned" the answer, and the tradeoffs were messy. As a mid-level programmer analyst, I didn't have the authority to decide. But I could help the decision emerge.

We had to design a system that would **dynamically prioritize deliveries**:

- Direct perishables to the stores with the highest demand,
- **Protect in-stock levels** across the broader network,
- And still fit within the scheduled delivery framework.

In one meeting, I reframed the issue away from "Which team's priority wins?" to "What balance of freshness, sales, and in-stock protection will best serve the customer?" By laying out the tradeoffs in a clear model, leaders could finally see the impact:

- If we sent more volume to high-demand stores but ignored others, we risked customer frustration in lower-volume locations.
- If we spread inventory evenly, we minimized risk but wasted product through spoilage.
- If we used a dynamic distribution approach — guided by demand signals and the existing delivery schedules — we could satisfy peak demand **and** maintain trust in the long tail of stores.

That framing unlocked the decision. The models I helped my leaders build would be used to influence executives. Executives stopped debating which silo should win and instead rallied around the system we built. The result was measurable: **on-shelf availability improved by 8% in perishables while spoilage decreased.**

I didn't have a VP title. I didn't control budgets or trucks. What I had was influence with my leaders — the ability to take competing priorities and reframe them into a balanced decision that served both business and customer.

The Three Dimensions of Influence

Together, these three create another flywheel: relationships get you heard, information makes you credible, and consistency makes you indispensable.

1. **Relational Influence**: Built on trust and connection. People know you, respect you, and are more likely to engage openly.
2. **Informational Influence**: Built on clarity. You bring the facts, trade-offs, and context that others need to make informed decisions.

3. **Positional Influence**: Built on consistency. Even without hierarchy, you become the default orchestrator because you keep the system moving.

Maxwell Application: The Law of Connection

In *The 21 Irrefutable Laws of Leadership*, John Maxwell writes: "Leaders touch a heart before they ask for a hand." This is the essence of influence.

In perishable replenishment, it wasn't just about numbers. It was about framing the impact on **customers** and **store associates**. When I connected the distribution challenge to a mother picking up bananas for her children, or to an associate frustrated by throwing away crates of strawberries, suddenly the debate wasn't abstract anymore. It was human.

That's the power of connection. Leaders follow those who can tie hard data to human outcomes. Influence doesn't start with a spreadsheet; it starts with empathy.

Walton Application: Influence Begins with Connection

Sam Walton modeled this long before analytics were commonplace. He was famous for walking the floor, talking directly to associates, and seeing the business through their eyes.

Walton understood that influence wasn't earned through data alone — it was built on connection. By valuing people first and results second, he created trust that made execution easier and faster.

For TPMs, the same principle holds true. You don't build lasting influence through spreadsheets or status decks; you build it by engaging with people, listening to their challenges, and showing them you care. When trust comes first, alignment and delivery follow naturally.

Saban Application: Raising the Standard

Nick Saban says: "Mediocre people don't like high achievers, and high achievers don't like mediocre people."

Influence without authority often means raising standards when others are comfortable with "good enough." TPMs who model excellence raise the bar for the whole team. Not everyone will welcome it at first — but clarity and conviction attract high performers. Influence grows when you consistently set the tone for what great looks like.

Framing: The TPM Superpower

Of all the skills that generate influence, **framing** is the most powerful.

TPMs constantly reframe chaos into clarity:

- Instead of "we have 15 blockers," you say, "there are three decisions we need to make."
- Instead of "engineering is behind," you say, "here's what will slip if we keep the current scope."
- Instead of "we don't know," you say, "here are two scenarios and what each implies."

Executives and stakeholders don't follow status updates — they follow frames that make decision-making easier.

Influence Without Manipulation

Influence is not about bending people to your will. Manipulation corrodes trust. True influence is service:

- You make work clearer and easier for others.
- You amplify the best ideas, regardless of origin.
- You guide the team toward outcomes they already want but can't yet see clearly.

When people realize your influence helps them succeed, they welcome it.

Key Takeaways

- TPMs don't rely on authority; they trade in influence.
- Influence is earned through relationships, information, and consistency.
- Framing is the superpower that transforms noise into clarity.
- Influence isn't manipulation — it's service that creates space for better decisions.

Chapter 3: Driving Decisions in the Gray Zone

"Leaders must be close enough to relate to others, but far enough ahead to motivate them."
— *John C. Maxwell*

"Commit to your business. Believe in it more than anybody else."
— *Sam Walton*

Living in Ambiguity

The hardest problems don't come with clean ownership charts. Who decides whether to prioritize customer features or technical debt? Who makes the call when compliance, product, and engineering disagree? These are the gray zones, areas with no obvious authority, unclear data, and competing priorities.

This is where TPMs prove their value. While others wait for perfect information or escalation, the TPM designs a path forward.

What the Gray Zone Looks Like

- **Multiple owners, no decider**: several leaders weigh in, but nobody commits.
- **Incomplete information**: data exists, but it's noisy, outdated, or contradictory.
- **Conflicting incentives**: teams optimize for their silo, not the system.
- **Decision fatigue**: leaders avoid the choice because every option carries risk.

Left unchecked, gray zones stall programs, burn out teams, and erode trust.

Field Story: Compliance vs. Customer Value

During one large-scale enterprise rollout, I sat in the middle of a standoff between **compliance, product, and engineering**.

- **Compliance** was demanding additional controls that would delay the launch by months.
- **Product** was pushing hard to meet a promised customer-facing date.
- **Engineering** argued that splitting their focus would break delivery momentum.

There was no single decision-maker. Each group was right in its own way — and deadlocked in conflict.

I didn't have the authority to decide. But I did have the responsibility to **help leaders make the decision.**

So, I worked directly with the **product team and business stakeholders** to calculate the costs and benefits of each path. Together, we built a simple decision framework:

> **Option 1:** Full compliance first → 3-month delay, higher cost, lower risk.
> **Option 2:** Partial compliance with mitigation → 4-week delay, medium cost, manageable risk.
> **Option 3:** Launch on time → no delay, lowest cost, highest risk (formally documented).

I then framed the decision for leadership:

- Here are the options.

- Here are the tradeoffs in cost, customer impact, and risk.
- Here's the deadline to decide if we want to keep the program on track.

The conversation shifted. Instead of endless debate, leaders had **structured tradeoffs with real numbers** to evaluate. Within days, they chose Option 2 — partial compliance with mitigations. The program moved forward, customers were served, and risk was transparently documented.

I wasn't the owner of the decision. Yet, by building the options in partnership with product and business — and framing them clearly for leadership — I worked with my partners and turned ambiguity into clarity and kept the program alive.

The TPM's Role: Decision Architect

In these situations, the TPM becomes a **decision architect**. Your job is not to make the decision but to design the environment where the decision can be made.

You do this by:

1. **Framing the Decision** — What exactly must be decided? What's at stake?
2. **Clarifying Owners** — Who is the decider? Who provides input? Who needs to be informed?
3. **Defining Tradeoffs** — Lay out the real options, their costs, and benefits.
4. **Setting a Deadline** — Ambiguity loves to linger; time-box it.
5. **Documenting the Outcome** — Capture the rationale so the team can move forward with alignment.

Maxwell Application: The Law of Navigation

John Maxwell teaches the *Law of Navigation*: "Anyone can steer the ship, but it takes a leader to chart the course."

That's what TPMs do in gray zones. You don't always hold the wheel, but you chart the course by mapping options, clarifying stakes, and ensuring the ship doesn't drift in circles.

Walton Application: Commitment Builds Confidence

Sam Walton captured the same spirit when he said: "Commit to your business. Believe in it more than anybody else."

For Walton, belief wasn't passive optimism, it was the fuel that carried him and his teams through uncertainty. His conviction gave others the courage to take risks and move forward.

The same is true for TPMs. When you step into the gray zone with clarity and commitment, you give others the confidence to act. Even when the path isn't obvious, your steady presence communicates: *"This matters. We can figure it out together."* That kind of conviction doesn't just move projects forward — it moves people forward.

The Power of Structured Ambiguity

Ambiguity isn't the enemy; unmanaged ambiguity is. By structuring the gray zone, TPMs transform confusion into choice. Leaders don't want you to tell them what to do. They want you to make it clear what doing nothing will cost.

Key Takeaways

- The gray zone is where TPMs earn disproportionate value.
- TPMs act as decision architects, not decision owners.
- Structure ambiguity with framing, ownership, tradeoffs, deadlines, and documentation.
- Leaders respect TPMs who turn paralysis into progress.

Part 2: Personal Growth & Resilience

"The greatest leader is not necessarily the one who does the greatest things. He is the one that gets people to do the greatest things."
— John C. Maxwell

Why Growth and Resilience Matter

Influence without authority is powerful, but it comes with a cost. TPMs sit at the intersection of competing demands, conflicting priorities, and constant change. If you're not careful, the very role that allows you to create impact can also drain your energy, blur your focus, and leave you burned out.

Anyone who has spent time leading large, complex programs knows that not every project ends the way you planned. Some miss deadlines. Others shift in scope until they hardly resemble the original vision. At times, despite the hours you pour in, initiatives stall or even fail outright. Those moments sting — but they also shape you.

John Maxwell reminds us in *Failing Forward* that "the difference between average people and achieving people is their perception of and response to failure." Failure isn't the end of leadership — it's the raw material for growth. Each difficult project leaves lessons about resilience, clarity, and the power of persistence.

Resilience and personal growth aren't luxuries — they're requirements. A TPM who cannot manage their own clarity, energy, and discipline will struggle to lead others. The best leaders build systems not only for their teams but also for themselves. They recognize that growth doesn't come in the easy seasons, but in the stretching, difficult ones — and resilience is forged in the fire of setbacks as much as in the glow of success.

The Hidden Work of the TPM

Every meeting, escalation, or late-night decision carries invisible weight. Teams look to you not just for updates but for steadiness. Executives expect you to absorb complexity and translate it into clarity. Colleagues depend on your ability to reframe problems without losing composure. This hidden work demands a strong inner core.

What This Part Will Cover

In this section, we'll shift from the external mechanics of TPM leadership to the inner disciplines that sustain it:

1. **From Firefighter to Architect** — moving beyond crisis response to building proactive systems.
2. **Resilience in the Face of Pressure** — practical habits for managing stress, setbacks, and relentless expectations.
3. **Energy, Fitness, and Focus** — how your personal well-being directly fuels your leadership capacity.

The Growth Flywheel

Just as TPMs design flywheels for programs, they must design one for themselves:

- **Identity** — knowing who you are and why you lead.
- **Habits** — building daily practices that reinforce clarity and resilience.
- **Reflection** — learning from experiences to grow stronger.

- **Renewal** — recharging physically, mentally, and emotionally to sustain performance.

When these elements reinforce each other, growth compounds — and so does your impact as a leader.

Setting the Stage

This part of the book is about strengthening the foundation that everything else rests on. Leadership begins with self-leadership. If you can manage your own energy, emotions, and habits, you can sustain the clarity and influence your teams depend on. **Growth isn't an accident, and resilience isn't innate; both are choices, built one practice at a time.**

Chapter 4: From Firefighter to Architect

"Pay attention to the basics. Small wins build confidence, and confidence builds resilience."
— Sam Walton

The Trap of Firefighting

Early in a TPM's career, it's easy to become the hero who solves problems in the moment: chasing down a dependency, escalating to leadership, or patching up last-minute risks. It feels productive, and sometimes it is. But firefighting as a way of life is unsustainable. Constant reaction keeps you busy, but it rarely moves the system forward.

Firefighting has three hidden costs:

1. **Exhaustion** — burnout from living in a cycle of urgency.
2. **Erosion of trust** — teams learn to expect chaos instead of clarity.
3. **Shallow impact** — fires get put out, but the root causes remain.

The Architect's Mindset

An *architect* TPM doesn't just solve today's fire, they design the conditions that prevent tomorrow's fire.

Instead of being reactive, they look at the system and ask:

- What process, cadence, or artifact could have prevented this?
- How can I make this decision easier next time?

- What can I build once that saves effort a hundred times?

The architect shifts from individual heroics to **systemic leverage**.

Tools of the TPM Architect

1. **Operating Cadences** — predictable forums for decisions, risk reviews, and alignment.
2. **Decision Maps** — clarity on who decides what, by when, eliminating endless debates.
3. **Risk Ledgers** — tracking risks transparently so surprises decrease.
4. **Metrics Dashboards** — the few signals that expose truth without noise.
5. **Narrative Templates** — reusable structures for telling the story at any level.

Each of these tools is a **force multiplier**: once created, it reduces firefighting downstream.

Field Story: The Launch That Stopped Burning

I remember when a high-stakes retail launch started as a daily inferno. Bugs surfaced, compliance raised last-minute red flags, and executives were calling meetings twice a day. The TPM shifted from fire suppression to architecture. They built a single **risk ledger** that captured every open risk, its owner, and mitigation status. They established a weekly decision cadence with executives, replacing ad-hoc escalations.

Within a month, the fires slowed. Not because the TPM worked harder, but because they designed a better operating system.

Maxwell Application: From Motion to Momentum

John Maxwell teaches that momentum is a leader's best friend, and momentum is rarely built in crisis. It's built through consistency and clarity that compound. When you install cadences and simple artifacts, you manufacture momentum: small wins, predictable progress, and rising confidence.

Walton Application: Small Wins, Big Confidence

Sam Walton had a bias for the basics, reminding leaders that *"small wins build confidence."*

He understood that momentum doesn't come from grand gestures but from simple, repeatable victories that prove progress is possible. Each win creates belief, and belief compounds into energy.

For TPMs, the lesson is clear: design simple wins into the system. Whether it's a quick dependency cleared, a risk log that brings order, or a decision framed clearly for executives, those early victories create confidence. And with confidence, teams execute faster, with less oversight, and with far greater unity.

Saban Application: The Pain of Discipline

Nick Saban often tells his players: "There are two pains in life. The pain of discipline and the pain of disappointment."

His point is simple — you either invest in discipline now or pay a bigger price later. That principle applies directly to program leadership.

When TPMs design alignment into systems — through clear decision maps, cadence rhythms, and simple wins — they are choosing the pain of discipline. It requires effort up front: consistency, preparation, and accountability.

But that discipline prevents the larger pain of disappointment — missed deadlines, rework, misalignment, and broken trust. Leaders who embrace discipline create confidence. And confidence creates momentum.

Making the Shift in Your Own Role

Ask yourself:

- Where am I still playing firefighter instead of architect?
- What artifact or cadence could eliminate this recurring fire?
- How can I spend one hour designing a system that saves my team dozens of hours later?

Key Takeaways

- Firefighting feels productive but drains energy and trust.
- The architect TPM designs systems that prevent chaos before it starts.
- Tools like cadences, decision maps, and risk ledgers shift the role from reactive to strategic.

- True leadership impact comes not from putting out fires, but from building environments where fires don't spread.

Chapter 5: Resilience in the Face of Pressure

"The difference between average people and achieving people is their perception of and response to failure."
— John C. Maxwell

"High expectations are the key to everything."
— Sam Walton

The Reality of Pressure

Every TPM faces seasons when the pressure is relentless: deadlines shift, leaders demand updates, teams stumble, and the program feels like it's teetering on the edge. In these moments, your resilience is tested. Pressure isn't a sign you're failing; it's a signal that what you're doing matters. But how you respond to that pressure determines whether you grow stronger or burn out.

Resilience Defined

Resilience is more than endurance. It's the capacity to **absorb stress, adapt quickly, and return with strength**. For TPMs, resilience means staying steady when the room is unsettled, keeping clarity when others are overwhelmed, and holding perspective when chaos narrows everyone's vision. Your steady calmness and well-placed passion are what move progress forward.

The Core Practices of Resilient TPMs

1. **Perspective Management** — They frame setbacks as lessons, not verdicts.
2. **Energy Regulation** — They understand when to push and when to recharge.
3. **Emotional Control** — They remain calm and credible under stress, which builds trust.
4. **Network Leverage** — They don't carry pressure alone; they involve the right partners early.

Practical Habits for Building Resilience

- **Breathe before you speak.** A pause gives you control over tone and presence.
- **Use micro-reflections.** After each tough meeting, jot down what worked, what didn't, and what to improve next time.
- **Guard your routines.** Sleep, workouts, and nutrition aren't optional—they're resilience anchors.
- **Name the real risk.** Anxiety multiplies in vagueness; define what's truly at stake to shrink the problem.
- **Keep a long lens.** Ask: *Will this matter in 12 weeks? 12 months? 12 years?* Most fires shrink under perspective.

Field Story: Holding Steady in a Crisis

During a large-scale system outage, executives flooded Slack with urgent questions while engineers scrambled to isolate the issue. The situation was primed for chaos: multiple conversations, conflicting updates, and no clear narrative for leadership.

The TPM stepped in and created order:

- **One source of truth:** They stood up a dedicated Slack channel for verified updates only.
- **Structured cadence:** They set up a Zoom bridge where engineering, product, and ops could align in real time. This ensured decisions weren't scattered across fragmented chats.
- **Delegated communication:** They aligned with the Operations team to send a consistent update to executive leadership every 30 minutes, eliminating duplicate pings and status noise.
- **Clear framing:** Every update to leadership included the known facts, current risks, and the next decision point.

The outage still took hours to resolve, but the program didn't spiral into panic. Executives felt informed, engineers stayed focused, and the broader organization trusted the process. Afterward, leaders noted to the TPM: *"You didn't just manage the outage, you managed us through it."* That's resilience in action, turning potential breakdown into trust-building leadership.

Maxwell Application: Failing Forward & the Rubber Band

John Maxwell often says in *Failing Forward*: "The difference between average people and achieving people is how they handle failure."

Resilient leaders treat setbacks as tuition — the price of learning. They don't fear mistakes; they fear missing the lesson.

Maxwell also teaches the *Law of the Rubber Band*: *"Growth stops when you lose the tension between where you are and where you could be."* Pressure is the tension that stretches us. If you embrace it, you grow. If you avoid it, you shrink.

Walton Application: Optimism Under Fire

Sam Walton believed optimism was a choice, even in the hardest moments. He said: *"High expectations are the key to everything."*

During tough seasons, Walton modeled resilience by walking stores, staying visible, and keeping expectations high. He didn't deny problems; he faced them with belief that solutions would emerge. TPMs who model this optimism shift the emotional tone of their programs.

Saban Application: Nothing is Given

Nick Saban once told his team: "Nothing is easy. Nothing is free. Nothing is given to you. You have to work for everything you get."

That mindset is at the heart of resilience. Leaders who expect ease are caught off guard by adversity. But leaders who embrace the reality that *nothing is given* aren't surprised when the pressure comes. They've already chosen discipline.

For TPMs, this means not waiting for perfect conditions or executive clarity. You earn influence, trust, and alignment by showing up every day with steadiness and grit.

The Calm Under Fire Routine

Resilience isn't accidental. You can train it. Here's a **five-step routine** TPMs can use in any high-pressure moment:

1. **Pause before responding.** Even a five-second breath resets your tone and presence.
2. **Name the reality.** State clearly: "Here's what we know, here's what we don't know." Anxiety multiplies in vagueness.
3. **Shrink the scope.** Identify the single next decision that matters most.
4. **Set a cadence.** Decide when the next update will be (e.g., "We'll regroup in 30 minutes"). Predictability builds trust.
5. **Stay human.** Thank the team, acknowledge the stress, and remind them of the mission.

Use this routine consistently, and people will begin to instinctively look to you when pressure spikes.

Turning Pressure into Growth

Resilience isn't avoiding stress — it's converting stress into strength. Every tough program builds capacity for the next one. By reflecting, adjusting, and recovering, TPMs become leaders people trust when the stakes are highest.

Key Takeaways

- Pressure is inevitable; your response shapes your impact.

- Resilience is built on perspective, energy, emotional control, and support networks.
- Habits like pausing, reflection, and routines protect you in the toughest moments.
- Calm presence under stress multiplies your influence and credibility.
- Resilient TPMs don't just survive pressure, they transform it into growth.

Chapter 6: Energy, Fitness, and Focus

"To succeed in business, you need to learn to manage your time, your energy, and your focus with the same discipline as your money."
— *Sam Walton*

Why Energy Matters for TPMs

Leadership isn't just mental. Every program you lead demands long hours, tough conversations, and sustained presence. When your energy is low, your influence shrinks. When you're physically strong and mentally sharp, you project confidence and steadiness that people instinctively follow.

Energy isn't a luxury — it's the foundation of resilience and clarity. TPMs who ignore their health eventually find their effectiveness eroded. **The best leaders treat energy as a strategic asset.**

The Energy Triad

TPMs can borrow from high-performance athletics: energy comes from the integration of **Body, Mind, and Focus**.

1. **Body** — Physical fitness, nutrition, and sleep are your baseline. A strong body handles stress better and recovers faster.
2. **Mind** — Mental resilience, reflection, and learning keep perspective sharp.
3. **Focus** — Directing attention with intention ensures energy is applied where it matters most.

When one leg of the triad collapses, the system tilts. When all three are reinforced, you become a leader who sustains momentum for others.

Field Story: The Marathon TPM

During one of the most demanding seasons of my career, I was leading a large-scale digital transformation project while training for the **Route 66 Marathon**. At first, it seemed impossible to balance. The project required constant alignment across multiple functions, while the marathon demanded disciplined preparation.

But I discovered that training didn't compete with my leadership; it fueled it. Early morning runs became my time to reflect on tough program challenges and mentally rehearse conversations I needed to lead. Long runs built not only physical endurance but also the patience to handle setbacks in the project. Choosing better nutrition for training carried over into sharper energy in afternoon executive meetings.

I found myself steadier, calmer, and clearer in decision-making. The lesson was simple: **physical training wasn't separate from leadership; it was a foundation for it.** By managing my body and mind well, I had more to give to my team and my company when the pressure was highest.

Building Your Focus Muscle

Focus is a skill, not just a trait. TPMs live in constant context-switching; Slack pings, meetings, escalations. To lead effectively, you must carve out space for **deep focus**:

- Schedule two "focus blocks" per week for strategic planning.
- Turn off notifications during key deliverables.
- Use journaling to reset priorities when noise overwhelms.

Focus is the multiplier: energy without focus is wasted effort. Focus without energy is burnout. Together, they drive impact.

Maxwell Application: The Law of Consistency

John Maxwell teaches in *The 15 Invaluable Laws of Growth*: "Motivation gets you going, but discipline keeps you growing."

Energy and focus aren't built in a day; they're built daily. Just as a runner trains mile by mile, leaders build stamina habit by habit. The TPM who treats workouts, reflection, and renewal as non-negotiable appointments doesn't just protect themselves, they multiply their influence by showing up consistently strong.

Walton Application: Daily Discipline Fuels Energy

Sam Walton once said: "I had to get up every day with my mind set on improving something."

That mindset wasn't about big, flashy moves. It was about the daily disciplines that compound over time. The same is true of energy and focus. Leaders who improve something small each day — whether it's their sleep, fitness, focus, or nutrition — create a reservoir of strength that others can depend on. While others burn out chasing quick wins, disciplined leaders sustain momentum by showing up consistently with energy and clarity.

Saban Application: Building Trust Through Consistency

Saban reminds us: "The difference between belief and trust can be the difference between a good team and a great team."

Belief is hoping your leader shows up. Trust is knowing they will. TPMs build trust not through one-time bursts of energy but through consistent habits — showing up with clarity, protecting priorities, and leading with focus. When your team can trust your consistency, they bring their best energy too.

The Energy Stack for Leaders

Think of your energy as a stack, built layer by layer:

1. **Sleep:** Guard 7–8 hours as seriously as a meeting. Recovery is strategy.
2. **Movement:** Train at least 4–5 times per week (strength + cardio). Fitness builds confidence and resilience.
3. **Nutrition:** Fuel for clarity, not convenience. Avoid sugar crashes before high-stakes meetings.
4. **Focus:** Carve out deep-work blocks (1–2 per week). Protect them from Slack and email.
5. **Renewal:** Practice reflection, journaling, or prayer. Renewal turns activity into wisdom.

Stack these consistently, and your presence in the room changes. People follow leaders who radiate calm, energy, and focus.

Reflection Exercise: Designing Your Energy System

Take 30 minutes this week to sketch your personal **Energy System**:

1. **Body:** What's one upgrade to your sleep, nutrition, or fitness routine you can commit to?
2. **Mind:** What practice (journaling, prayer, reading) will sharpen your perspective?
3. **Focus:** Where can you schedule 2 hours of uninterrupted deep work this week?
4. **Renewal:** What activity helps you recharge? How will you protect time for it?

Write your answers on a single page and post it where you'll see it daily. Energy isn't managed by hope — it's managed by design.

Key Takeaways

- Energy is not optional — it's a leader's most important asset.
- Physical fitness, mental resilience, and focus form the triad that sustains performance.
- Daily habits (movement, sleep, nutrition, deep work) protect clarity under stress.
- Leaders who manage energy and focus well multiply their influence and impact.

Part 3: Scaling Your Impact

"Outstanding leaders go out of their way to boost the self-esteem of their personnel. If people believe in themselves, it's amazing what they can accomplish."
— Sam Walton

From Individual Contributor to Force Multiplier

Up to this point, we've focused on leading without authority and building resilience within yourself. But leadership at scale requires more than personal clarity and energy — it requires the ability to amplify impact across entire organizations.

Scaling impact as a TPM means shifting from being the go-to problem solver to becoming the architect of systems, cadences, and narratives that allow dozens or even hundreds of people to move faster together. It's one thing to solve a thorny dependency for your team. It's another thing entirely to design a process that prevents that dependency from slowing down *any* team again.

This is where many TPMs either plateau or break through. Some continue to operate as highly effective individual contributors, carrying projects on their shoulders. Others recognize that true leadership is about multiplication, not addition. John Maxwell puts it plainly in *The 21 Irrefutable Laws of Leadership*: *"To add growth, lead followers. To multiply, lead leaders."*

The force-multiplying TPM understands that their job is not just to clear the path for one project, but to equip people, refine systems, and create rhythms that outlast their direct involvement. They move from being firefighters to architects, from coordinators to catalysts. And in doing so, they unlock an exponential effect: projects align more naturally, decisions are made more quickly, and teams operate with a level of clarity that accelerates the whole enterprise.

Force multiplication doesn't come from working harder — it comes from leading smarter. It's the pivot from asking, *"How can I solve this?"* to *"How can I design a way for others to solve this faster?"* That shift is the mark of a leader whose influence scales far beyond their personal capacity.

Why Scaling Matters

As programs grow in complexity, your influence cannot depend solely on your personal presence in every meeting. If everything relies on you being in the room, you'll eventually become the bottleneck. Scaling is about building the tools, rituals, and frameworks that let clarity and alignment flow even when you're not there.

What This Part Will Cover

In this section, we'll explore the practices that allow TPMs to extend their influence beyond one program or one team:

1. **From Executor to Strategist** — how to move beyond delivery tracking into shaping business outcomes.
2. **The TPM as Force Multiplier** — using data, AI, and systems thinking to scale impact across an organization.
3. **Building Your Leadership Brand** — crafting the visibility, trust, and reputation that accelerate your career and expand your influence.

The Leadership Multiplier Effect

Think of this part as designing your personal *multiplier effect*:

- **Systems** that replace heroics with repeatable clarity.
- **Data and technology** that extend your insights further, faster.
- **A leadership brand** that ensures people call you into the right rooms at the right time.

When you scale your impact, you stop being just the person who "keeps projects on track" — you become the leader executives trust to guide transformation.

Chapter 7: From Executor to Strategist

"A leader is one who sees more than others see, who sees farther than others see, and who sees before others do."
— *John C. Maxwell*

The Executor Trap

Many TPMs begin their careers focused on execution: tracking milestones, updating dashboards, escalating blockers. These are valuable contributions, but they can create a ceiling. If your role is defined only by execution, you risk being seen as a coordinator rather than a leader.

Execution ensures the trains run on time. Strategy determines **whether the trains are headed in the right direction.**

The Shift to Strategic Leadership

A strategic TPM doesn't abandon execution, they elevate it. They connect delivery to outcomes. They ask not just *"Are we on track?"* but *"Are we delivering what matters most?"*

Strategists operate in three dimensions:

1. **Business Alignment** — They understand revenue, cost, customer value, and regulatory context.
2. **Systems Thinking** — They see interdependencies across teams, platforms, and roadmaps.
3. **Decision Framing** — They guide other leaders to make tradeoffs that balance speed, quality, and impact.

Practical Shifts You Can Make

Making the leap from executor to strategist doesn't happen overnight. It's built on a series of small but deliberate shifts in how you think, how you speak, and how you frame your work.

1. Stop reporting activity; start reporting outcomes.
   a. Executor update: "Team A completed 25 stories this sprint."
   b. Strategist update: *"Team A reduced checkout latency by 12%, which directly improved conversion and revenue lift."*
      Outcomes link execution to measurable value — which is what executives care about.

2. Reframe metrics to speak the language of the business.
   a. Executor metrics: story points, defects resolved, sprint velocity.
   b. Strategist metrics: customer adoption, cost savings, reduced downtime, regulatory risk avoided.
      TPMs must translate delivery signals into metrics that tell the business story.

3. **Ask higher-level questions.**
   Executors ask, *"Are we on track?"* Strategists ask:
   a. "What business problem are we solving?"
   b. "What decision needs to be made this quarter?"
   c. "What's the opportunity cost if this slips?"
      These questions elevate the conversation from task tracking to impact framing.

4. **Connect dots across silos.**
   Executors focus on their single program. Strategists show

how multiple initiatives interact. For example: *"Feature A from Product ties directly into Infrastructure's roadmap, and both depend on Data Services' upgrade. If we align sequencing, we save three months."* This cross-view creates leverage.

5. **Anticipate tradeoffs before escalation.**
   Instead of surfacing a blocker and waiting for leadership to decide, strategists present options:
   - Option 1: Launch now with limited functionality — gain speed but incur rework later.
   - Option 2: Delay by 6 weeks to implement scalable architecture — higher cost now, long-term savings. This positions you as a thought partner, not just a messenger.

6. **Craft narratives instead of status dumps.**
   Executors provide updates that list progress and blockers. Strategists synthesize:
   a. "Here's the outcome we're driving, here's where we are, here are the risks and tradeoffs, and here's my recommendation."
   Leaders follow narratives; they skim past lists.

Field Story: The Pharmacy Delivery Program

When working on securing engineering leadership commitments for a large-scale **pharmacy delivery rollout**, I initially focused on execution, aligning engineers, and discussing feature deadlines. But the breakthrough came when our product team and I reframed the discussions not around timelines but around **customer access**: how many patients would be served, how delivery improved adherence, how the program reduced costs for both the business and consumers.

By shifting from executor to strategist, we influenced executive investment decisions, helped the business and product team secure additional resources from partner teams, and positioned the program not as "a tech launch," but as **a healthcare transformation initiative.**

Maxwell Application: The Law of the Lid

In The 21 Irrefutable Laws of Leadership, Maxwell writes: "Leadership ability is the lid that determines a person's level of effectiveness."

If you operate only as an executor, you cap your influence. But when you rise into strategy — linking delivery to business outcomes, anticipating ripple effects, shaping decisions — you lift your lid.

Maxwell also teaches the *Law of the Big Picture*: leaders don't just see the next task; they see how it connects to the mission. Strategist TPMs live in that big picture.

Walton Application: Customers First, Always

Sam Walton was obsessed with connecting strategy to customer value. He constantly asked, *"How does this reduce costs for the customer? How does it improve their experience?"*

For TPMs, this means you don't just report delivery metrics (velocity, story points). You translate them into **customer outcomes** (faster checkout, lower spoilage, better access to healthcare). When you speak Walton's language of value and savings, executives hear strategy.

Building Strategic Credibility

To be seen as a strategist, you must:

- **Speak the language of executives.** Link program updates to revenue, customer value, and risk.
- **Anticipate the second-order effects.** How will today's decision ripple six months from now?
- **Bring solutions, not just problems.** Executives already know the risks; they want your framing of the options.

Key Takeaways

- The executor ensures delivery; the strategist ensures impact.
- TPMs must elevate conversations from activity to outcomes.
- Strategic TPMs frame decisions, align business goals, and anticipate ripple effects.
- Moving from executor to strategist is the inflection point that shifts TPMs into true leadership.

Chapter 8: The TPM as Force Multiplier

"Control your expenses better than your competition. This is where you can always find the competitive advantage."
— Sam Walton

"Leaders who develop leaders multiply growth. Leaders who develop followers only add to it."
— John C. Maxwell

From Individual Impact to Organizational Leverage

The best TPMs don't just manage programs; they build frameworks that allow **entire organizations** to move faster, with less friction, and more alignment. This is what it means to be a force multiplier. Your actions and tools amplify the productivity, clarity, and decision-making of everyone around you.

What Does a Force Multiplier Look Like?

Force multipliers are not about working harder, they're about designing systems that make *everyone* more effective.

- A decision framework that clarifies ownership for dozens of teams.
- A system dashboard that reduces status churn across hundreds of stakeholders.
- An AI-assisted narrative that distills thousands of data points into a two-page executive update.
- A cadence that eliminates redundant meetings while keeping leaders aligned.

Force multipliers are not about working harder, they're about designing systems that make *everyone* more effective.

The Three Levers of Multiplication

1. Systems and Cadences
   a. Executors run meetings; force multipliers design operating rhythms.
   b. Weekly decision forums, risk reviews, and dependency syncs that prevent chaos before it spreads.
2. Data and Visibility
   a. Executors report metrics; force multipliers design dashboards.
   b. By curating the vital few metrics and automating their visibility, TPMs create shared truth that reduces debate and speeds decisions.
3. Technology and AI
   a. Executors take notes; force multipliers train AI to summarize, synthesize, and draft.
   b. Modern TPMs leverage AI to pull insights from Slack, Jira, Confluence, and email into coherent risk reports or decision briefs — freeing them to focus on judgment and framing.

Field Story: Multiplying Clarity at Scale

In one global program, teams across six work locations, across time zones, were struggling with misalignment. Each region reported its status differently, which frustrated executives, and engineering leads spent hours reconciling the reports.

The TPM designed a **single dashboard** that pulled data from all regions into a unified view of outcomes, risks, and dependencies. They paired this with a **bi-weekly decision cadence** that replaced four separate update meetings. To amplify further, they used an **AI assistant** to generate executive summaries, saving hours of manual work.

The result? Faster decisions, less meeting fatigue, and trust restored at the executive level. The TPM's personal hours didn't multiply, their impact did.

Maxwell Application: The Law of Explosive Growth

In *The 21 Irrefutable Laws of Leadership*, John Maxwell writes: "Leaders who develop followers add growth. Leaders who develop leaders multiply growth."

Force multiplication is leadership at scale. You're not just managing your own clarity; you're designing systems, cadences, and artifacts that **develop clarity in others**. That's how growth compounds.

Walton Application: Systems as Culture

Sam Walton believed Walmart's success wasn't just about prices, but about **disciplined systems**: controlling expenses, standardizing processes, and multiplying scale.

TPMs can borrow this mindset. Your systems — risk ledgers, decision maps, cadences — aren't "admin overhead." They're culture. They teach people how to operate with clarity and speed.

Becoming the Multiplier in Your Org

To step into this role:

- Ask: "What system could I build once that would save everyone hours weekly?"
- Use AI as your junior analyst — draft the narrative, then apply your judgment.
- Train leaders to rely on frameworks, not fire drills.
- Share playbooks, not just updates — create templates others can reuse.

Key Takeaways

- Force multipliers design systems, not just meetings.
- They curate visibility, turning metrics into clarity.
- They leverage technology and AI to scale insight.
- Their success is measured not by what they deliver alone, but by how much faster and clearer the organization becomes around them.

Chapter 9: Building Your Leadership Brand

"People buy into the leader before they buy into the vision."
— *John C. Maxwell*

"There's absolutely no limit to what plain, ordinary, working people can accomplish if they're given the opportunity and encouragement to do their best."
— *Sam Walton*

Why Your Brand Matters

In large organizations, decisions about opportunities, promotions, and trust are often made in rooms where you're not present. In those moments, your **leadership brand** speaks for you. It's not about self-promotion; it's about the consistent impression you create through your credibility, clarity, and the value you bring.

A strong TPM leadership brand ensures that when leaders think of who can handle complexity, drive clarity, or lead transformation, **your name comes up.**

What Makes a Leadership Brand?

- **Credibility** — You deliver what you promise, consistently.
- **Clarity** — You make complex things simple and help others see the path forward.
- **Contribution** — You're known for adding value, not noise.
- **Character** — You lead with integrity, even when pressure is high.

Practical Ways to Build Your Brand

- **Craft the right narrative.** Don't just say *what* you did. Explain *why it mattered.* Frame results in terms of outcomes, not activity.
- **Be visible in the right rooms.** Volunteer for cross-functional initiatives, present updates at leadership forums, and make sure executives see you synthesizing complexity into clarity.
- **Invest in communication.** Clear writing and structured storytelling build confidence in your leadership. A two-slide narrative can shape more trust than a twenty-slide deck.
- **Mentor and elevate others.** Leaders who develop others multiply their influence and build a reputation for generosity.
- **Use artifacts as brand builders.** A well-designed decision map, dashboard, or program charter can live beyond you, carrying your signature style of clarity.

Field Story: From Invisible to Indispensable

In my early Walmart career, I was dependable but overlooked. My updates were thorough, but they blended into the noise.

The shift came when I started presenting differently. Instead of reporting tasks, I began framing updates as **strategic narratives**:

- "Here's the outcome we're driving."
- "Here's where we are against that outcome."
- "Here are the top risks and tradeoffs."
- "Here's my recommendation."

Within weeks, leaders began noticing. Within months, executives requested my updates by name. I didn't ask for visibility; my **brand of clarity created it**.

When a new transformation program launched, I was tapped to lead. Not because I raised my hand, but because my brand had already done the talking.

Maxwell Application: The Law of Buy-In

John Maxwell teaches: "People buy into the leader before they buy into the vision."

Your updates, your artifacts, your presence — they're all opportunities for people to buy into you. If they trust your clarity and credibility, they'll follow your recommendations.

Maxwell also teaches the *Law of Solid Ground*: trust is the foundation of leadership. Your brand isn't built in a single meeting; it's reinforced in every small commitment kept and every honest framing of risk.

Walton Application: Extraordinary in the Ordinary

Sam Walton knew that talking with associates and listening to customers firsthand built a brand of **accessibility and authenticity**. He believed that caring for associates was the surest way to build long-term success.

He believed Walmart's greatest strength wasn't systems or stores — it was people. He said:

"There's absolutely no limit to what plain, ordinary, working people can accomplish if they're given the opportunity and encouragement to do their best."

Your leadership brand isn't built on deliverables. It's built on the belief you place in others. TPMs who empower teams, celebrate contributions, and create opportunities unleash extraordinary results from ordinary people. That's the mark of leadership that lasts.

TPMs can mirror this by:

- Being visible in cross-functional forums.
- Engaging directly with engineers, product managers, and ops.
- Showing consistency in how you show up.

Brand isn't what you say about yourself — it's what others consistently experience when they work with you.

Saban Application: Culture Defines You

Nick Saban teaches: "Culture is what you do every day. We do what we do because that's who we are."

Your leadership brand isn't a slide deck — it's your daily actions. Just as Saban's teams know the culture of discipline and excellence through daily practice, your colleagues recognize your brand through your consistent leadership in meetings, updates, and the development of others. Culture is brand. And a brand is built every day.

Sustaining Your Brand

Your leadership brand isn't built in a single meeting; it's reinforced in every interaction. Every email, meeting, artifact, and escalation either strengthens or weakens it. Consistency matters more than flash. When people know what to expect from you, steady clarity, trustworthy framing, thoughtful perspective, they will follow your lead.

Key Takeaways

- Your leadership brand is the story people tell about you when you're not in the room.
- Strong brands are built on credibility, clarity, contribution, and character.
- Communication, visibility, and mentoring are practical levers to strengthen your brand.
- Artifacts can extend your influence beyond the meetings you attend.
- A consistent brand ensures you're seen not just as an executor, but as a leader others trust with transformation.

Reflection Prompts: Building Your Leadership Brand

Use these questions as a personal checkpoint. Set aside 15–20 minutes to write your answers. The more honest you are, the more valuable this exercise will be:

1. Current Brand Reality
   a. If three colleagues were asked to describe me in three words, what would they say?
   b. Do I agree with that description? Why or why not?
2. Desired Brand Identity
   a. What three words do I *want* people to use when describing me as a leader?
   b. What actions or habits would reinforce those words?
3. Gaps to Close
   a. Where do my current actions undermine the brand I want to build?
   b. What recurring behaviors could I change in the next 30 days to strengthen my leadership presence?
4. Artifacts and Visibility
   a. What artifacts (dashboards, narratives, frameworks) could I create that carry my brand of clarity even when I'm not in the room?
   b. Where do I need to show up more visibly (leadership meetings, cross-functional forums, mentoring sessions)?
5. Long-Term Legacy
   a. Five years from now, what do I want my leadership brand to be known for in this organization or industry?
   b. What decisions today will move me toward that legacy?

Part 4: Future-Proofing Your Career

"Change is inevitable. Growth is optional."
— *John C. Maxwell*

Why Future-Proofing Matters

The role of TPM is evolving faster than ever. AI is reshaping how we gather data and summarize updates. Organizations are flattening hierarchies and demanding faster decisions. Programs are larger, more global, and more complex. What made you effective yesterday won't be enough tomorrow.

Future-proofing your career is about building the adaptability, skills, and perspective that will keep you relevant and valuable in any environment. It's about preparing not just to survive change, but to lead through it.

From Program Leader to Enterprise Influencer

The TPM of the future is more than a delivery expert. They are a **strategic integrator** — the leader who sees across silos, translates complexity into clarity, and guides decisions that shape the enterprise. Future-proofing means shifting your mindset from managing the present to anticipating and shaping the future.

What This Part Will Cover

In this section, we'll explore how to set yourself apart as the future unfolds:

1. **The AI-Accelerated TPM** — leveraging emerging tools to expand your impact rather than being replaced by them.
2. **From TPM to Executive** — mapping the path from program leadership into broader organizational leadership roles.
3. **Beyond Corporate** — exploring how TPM skills translate into entrepreneurship, consulting, and coaching.

Your Career as a System

Think of your career the way you think of a program: a system with inputs, outputs, dependencies, and opportunities for leverage. **By investing in your adaptability, expanding your influence, and clarifying your long-term purpose, you design a career that doesn't just react to change; it thrives on it.**

Chapter 10: The AI-Accelerated TPM

"The pessimist complains about the wind. The optimist expects it to change. The leader adjusts the sails."
— *John C. Maxwell*

Why AI Matters for TPMs

Artificial Intelligence is no longer a buzzword — it's becoming embedded in the daily fabric of how teams communicate, make decisions, and deliver work. For TPMs, this shift is especially significant. Much of the repetitive work we once owned — compiling notes, gathering status, building dashboards — can now be automated. That doesn't diminish the TPM role; it **elevates it**.

The TPM who learns to harness AI as a partner moves from being a coordinator to becoming a **strategic force multiplier.**

Where AI Can Accelerate You

I've heard it said that AI won't replace the TPM, but the TPM that uses AI will. Use these tips to accelerate your impact.

1. **Information Synthesis**
   a. Drafting meeting summaries from Zoom transcripts.
   b. Pulling risks and dependencies from Jira, Confluence, or Slack.
   c. Distilling hundreds of messages into a one-page executive brief.

2. **Decision Support**

    a. Running "what-if" scenarios across multiple data sources.
    b. Highlighting likely failure modes in a roadmap.
    c. Suggesting tradeoffs based on historical patterns.

3. **Productivity Leverage**
    a. Automating repetitive tasks like report formatting or follow-up reminders.
    b. Creating dashboards that update in real time.
    c. Drafting first versions of program charters, risk ledgers, or decision maps.

The Human Advantage

AI can draft, summarize, and analyze. But only humans can:

- **Build trust.** Relationships are not automated.
- **Make judgment calls.** Tradeoffs require values and priorities.
- **Tell the story.** Executives need framing, not just data.
- **Lead through crisis.** Calm presence under pressure is uniquely human.

The future TPM doesn't compete with AI; they **direct it** — using it to handle the noise while they focus on clarity, influence, and leadership.

Field Story: AI as Your Junior Analyst

With a global portfolio, I face the challenge of synthesizing updates from five time zones, dozens of Slack channels, emails, Confluence, and hundreds of Jira tickets. Instead of burning out, I use an AI assistant to:

- Summarize all updates into a single weekly digest.
- Flag high-risk dependencies automatically.
- Draft the first pass of the weekly executive report.

I then review, refine, and frame the narrative. The result? Executives receive clearer updates in half the time, engineers have fewer interruptions, and I am more able to focus on strategic alignment. AI hasn't replaced me; it gave me and my TPMs leverage.

Practical Steps to Become an AI-Accelerated TPM

- **Audit your workload.** Identify tasks that are repetitive, data-heavy, or easily templated. Automate those first.
- **Experiment small.** Use AI to draft meeting notes or risk logs before scaling up.
- **Pair judgment with automation.** Treat AI as your junior analyst — it drafts, you refine.
- **Stay curious.** The tools are evolving; commit to continuous learning.

Maxwell Application: The Law of Timing

In *The 21 Irrefutable Laws of Leadership*, Maxwell says: "When to lead is as important as what to do and where to go."

Adopting AI now is a timing play. Leaders who wait until tools are "perfect" will fall behind. TPMs who act early — experimenting, adapting, learning — will set the pace for their organizations.

Maxwell's *Law of the Lid* also applies; leadership ability determines effectiveness. AI lifts that lid by handling lower-value tasks, freeing you to lead at a higher level.

Walton Application: Efficiency is Advantage

Sam Walton built Walmart on operational efficiency — controlling costs, streamlining logistics, and standardizing systems. AI is the modern continuation of that principle.

Just as Sam insisted on real-time visibility into inventory and sales, TPMs should insist on real-time clarity through AI. Leaders who harness these efficiencies free capacity for innovation and customer value.

Key Takeaways

- AI is not a threat to TPMs — it's an accelerator for those who adapt.
- Automate the repetitive, data-heavy work so you can focus on leadership.
- Use AI to multiply your reach, not replace your judgment.
- The TPM of the future is part strategist, part system designer, and part AI conductor.

Quick Wins: 5 Ways to Use AI as a TPM This Month

You don't need a massive overhaul to start leveraging AI. Begin small and build momentum:

1. Meeting Summaries
   a. Feed Zoom or Teams transcripts into an AI tool.
   b. Get concise summaries with action items, then circulate within 15 minutes.
2. Risk & Dependency Extraction
   a. Paste backlog items, Jira tickets, or Slack updates into AI.
   b. Ask: "What risks or cross-team dependencies are hidden here?"
3. Executive Report Drafting
   a. Provide AI with last week's updates and ask for a 1-page narrative.
   b. Focus your time on refining and framing instead of starting from scratch.
4. Decision Options Framing
   a. Input constraints, goals, and risks.
   b. Ask AI: "What are three tradeoff scenarios with pros/cons?"
   c. Use as input for your leadership decision forums.
5. Artifact Templates
   a. Ask AI to generate first drafts of decision maps, program charters, or risk ledgers.
   b. Standardize and reuse across initiatives.

Action: Choose one of these quick wins and implement it this month. Treat AI as your **junior analyst** — it drafts, you decide.

Chapter 11: From TPM to Executive

"Leaders become great not because of their power, but because of their ability to empower others."
— *John C. Maxwell*

The Inflection Point

At some stage in your career, the question shifts from *"How do I deliver programs successfully?"* to *"How do I shape the organization itself?"* This is the inflection point between TPM and executive. Many TPMs plateau at being great orchestrators of delivery — but the path to executive leadership requires broader vision, influence, and the ability to guide strategy, not just execution.

What Changes at the Executive Level

- **From Programs to Portfolios:** Executives don't just oversee one transformation; they manage dozens, balancing resources and priorities across the enterprise.
- **From Delivery to Vision:** Success is measured less by meeting deadlines and more by shaping direction, culture, and long-term outcomes.
- **From Detail to Narrative:** Executives can't live in Jira tickets or sprint boards. They need a story of where the organization is headed and why it matters.
- **From Coordination to Empowerment:** Rather than driving every decision, executives build leaders who can make decisions without them.

The Skills to Develop on the Path

1. **Strategic Thinking** — Learn to see beyond milestones into market dynamics, customer value, and enterprise-level tradeoffs.
2. **Financial Acumen** — Executives live in numbers: budgets, ROI, operating margin. The TPM who speaks this language gets a seat at the table.
3. **Influence Across the Enterprise** — Move from cross-functional influence to enterprise influence, shaping decisions that ripple across entire business units.
4. **Executive Communication** — Learn to tell stories that simplify complexity for the C-Suite and boardroom. Two slides can win more trust than twenty when framed with clarity.
5. **Talent Development** — Executives aren't judged on what they deliver alone, but on the leaders they grow with them.

Field Story: The Transition to Portfolio Leadership

When I became Senior Manager of Data & Analytics at Walmart, the scale of my role shifted dramatically. I wasn't just guiding one cross-functional program anymore. I was shaping how multiple initiatives connected: data platforms, analytics tools, and enterprise reporting.

The turning point came when I stopped reporting activity and started **framing strategy**. Instead of saying, "The data warehouse project is 70% complete," I began asking:

- "How does this investment reduce cost to serve?"
- "How will this system unlock customer insights that drive revenue?"

- "What talent do we need to grow so these platforms sustain beyond me?"

That pivot moved me from executor → strategist → future executive material. Eventually, as a Director of TPM for Health & Wellness, I wasn't measured by my own delivery. I was measured by how many leaders I empowered, how much trust I built at the C-Suite, and how confidently I helped my peers shape Walmart's healthcare vision.

Executives trusted me not because I controlled details, but because I **framed vision, developed people, and multiplied outcomes.**

Maxwell Application: The Law of the Lid & Law of Legacy

John Maxwell's *Law of the Lid* says: "Leadership ability is the lid that determines a person's level of effectiveness."

Your execution can only take you so far. To rise into executive leadership, you must raise your leadership lid, expanding influence, business fluency, and people development.

Maxwell also teaches the *Law of Legacy*: leaders live beyond their projects. Executives aren't remembered for Jira boards or milestone charts. They're remembered for the culture they shaped, the leaders they grew, and the transformation they guided.

Walton Application: Empower People, Multiply Impact

Sam Walton modeled this long before leadership theory made it popular. He built Walmart by empowering associates, giving them ownership, and boosting their confidence. He knew **frontline empowerment scaled faster than executive control.**

For TPMs aspiring to executives, the same principle applies, you don't rise by controlling more, you rise by equipping more people to lead.

Practical Steps Toward Executive Roles

- Volunteer for initiatives with **enterprise impact** (e.g., cost-saving programs, market-entry projects).
- Partner with Finance or Strategy teams to deepen your **business fluency.**
- Mentor at least two future TPM leaders — demonstrate that you can scale leadership.
- Learn the **language of the boardroom**: risk, ROI, long-term growth.
- Create one **executive-level artifact** (vision narrative, investment proposal, strategy paper) every quarter.

Key Takeaways

- Moving from TPM to executive means shifting from delivery expert to enterprise leader.
- The transition requires skills in strategy, finance, influence, communication, and talent development.
- Executives are not measured by what they control, but by what they empower others to achieve.
- The best way to be chosen for an executive role is to start **operating like one now.**

Chapter 12: Beyond Corporate

"The greatest mistake we make is living in constant fear that we will make one."
— John C. Maxwell

Why Look Beyond?

For many TPMs, the corporate world is where the journey begins —
but it doesn't have to be where it ends. The skills that make you
effective inside large organizations — orchestrating clarity, aligning
diverse stakeholders, scaling systems — are highly valuable outside of
corporate structures. Beyond the org chart, these skills can fuel
entrepreneurship, consulting, or coaching careers where you own
your time, your impact, and your legacy.

The Transferable Skills of a TPM

- **Influence without authority** → Key in consulting
  engagements, where trust must be earned quickly.
- **Systems thinking** → Essential in entrepreneurship, where
  you're designing processes from scratch.
- **Communication and storytelling** → Core to coaching and
  advisory roles, where clarity accelerates others' growth.
- **Resilience under pressure** → Invaluable when building
  something new, where resources are scarce and uncertainty is
  high.

Options Beyond Corporate

1. Entrepreneurship
    a. Build products, services, or platforms that solve problems you've seen firsthand.
    b. Use your experience in orchestrating programs to design scalable businesses.
2. Consulting
    a. Offer your expertise in program design, digital transformation, or change management.
    b. Help organizations avoid the pitfalls you've navigated countless times.
3. Coaching and Mentoring
    a. Leverage your leadership journey to guide others.
    b. Focus on developing the next generation of TPMs, product leaders, or executives.
4. Portfolio Careers
    a. Combine writing, speaking, advisory work, and part-time leadership into a mix that maximizes freedom and impact.

Field Story: Raising My Lid Inside Corporate

In recent years, I've been intentional about not just leading programs, but **growing people**. At Walmart, I've led digital transformations and complex health initiatives — but alongside that work, I've also begun to apply the principles I learned from the **Maxwell Leadership Team** to mentor TPMs and rising leaders around me.

That shift has been transformational. My impact is no longer measured just in delivery outcomes. It's measured in the **growth of others**:

- A peer who gained confidence presenting to executives because I coached them through narrative framing.
- A junior TPM who learned to design a risk ledger and now leads programs with steadier confidence.
- Cross-functional leaders who now think in systems because I shared a flywheel approach with them.

Inside Walmart, I've realized my greatest impact isn't just in the systems I build — **it's in the people I help grow**.

Maxwell Application: The Law of the Rubber Band

John Maxwell teaches in *The 15 Invaluable Laws of Growth*: "Growth stops when you lose the tension between where you are and where you could be." A rubber band is useless unless it is stretched.

That tension is the spark of entrepreneurship. It's the restless pull that asks: What else is possible? How can I serve more? Where can I expand my influence?

For TPMs, the temptation is to stay in the comfort zone of execution. But growth — and true legacy — comes when you stretch beyond familiar systems and step into the unknown. Whether that's mentoring others, coaching, writing, or building something new, the stretch creates capacity you didn't know you had.

Entrepreneurial spirit isn't reckless. It's the discipline to lean into that tension with courage, knowing that the risk of staying comfortable is greater than the risk of stretching.

Walton Application: Experiment and Empower

Sam Walton built Walmart by empowering people and experimenting constantly. He didn't just think about what worked today; he tested what might work tomorrow.

That same entrepreneurial spirit applies inside corporate life. You don't need to quit to think "beyond corporate." You can mentor, experiment, and test new leadership approaches inside your current role — while preparing for how those same skills could translate into speaking, teaching, or coaching down the road.

Designing Your Own Beyond-Corporate Path

- **Clarify your purpose.** What problems do you want to solve? What people do you want to help?
- **Test small.** Start a side coaching or advisory project before leaving corporate entirely.
- **Build visibility.** Publish insights, frameworks, or case studies to show your expertise beyond your employer.
- **Plan financially.** Freedom comes not only from opportunities but from preparation.

Key Takeaways

- The TPM skillset translates directly into entrepreneurship, consulting, and coaching.
- Influence, systems thinking, and resilience are valuable in any environment.

- Beyond corporate, you own your time, your impact, and your legacy.
- You don't have to wait — start building your "beyond corporate" options now.

Exercise: Design Your Beyond-Corporate Experiment

Before you leap, you can *test* your future. Use this simple framework to explore what "beyond corporate" could look like for you:

1. Clarify Your Why
   a. What excites you about the idea of entrepreneurship, consulting, or coaching?
   b. Write one sentence that captures your motivation.
2. Identify a Problem You Can Solve
   a. What problem do you see repeatedly in organizations that you could solve faster, cheaper, or better?
   b. Write down three examples.
3. Package Your Expertise
   a. What artifact, framework, or playbook do you already use that others would pay for?
   b. Could it become a workshop, a coaching program, or a consulting offer?
4. Test Small
   a. Offer a free workshop, take on one coaching client, or publish a framework online.
   b. Measure the response: Did people engage? Did they ask for more?
5. Plan Your Next Step
   a. Based on your test, what is one action you can take in the next 30 days to move your beyond-corporate vision forward?

Action Prompt: Write your answers in a journal or a one-page document. By externalizing your thoughts, you'll begin turning a dream into a plan, and a plan into a reality.

Conclusion — Leading Change, Living Growth

"A leader is one who knows the way, goes the way, and shows the way."
— *John C. Maxwell*

The Journey Beyond Delivery

You began this book as a TPM who may have felt caught in the middle — accountable for outcomes without the authority to control them. Along the way, we've explored how to turn that paradox into power: building influence without authority, cultivating resilience, scaling your impact, and preparing for the future.

The lesson is clear: **true TPM leadership goes far beyond managing delivery.** It's about shaping the system, multiplying the effectiveness of those around you, and leading with presence and clarity when others are paralyzed by ambiguity.

The Core Commitments

As you move forward, remember the four commitments that will define your growth:

1. **Clarity** — Always reframe chaos into decisions, tradeoffs, and next steps.
2. **Influence** — Build trust, credibility, and consistency so your presence carries weight.
3. **Resilience** — Protect your energy, routines, and perspective to sustain leadership.
4. **Growth** — Keep scaling your skills, systems, and vision so you're always moving forward.

Your Next Chapter

Whether you remain in corporate life, rise to the executive ranks, or chart a path beyond into entrepreneurship or coaching, the same principles apply:

- Lead with clarity.
- Serve with influence.
- Stand with resilience.
- Pursue growth with intention.

The best TPMs don't just deliver outcomes. They deliver transformation — for their teams, their organizations, and themselves.

The Maxwell Charge

John Maxwell reminds us that leadership is influence — nothing more, nothing less. But influence is not static; it grows as you grow. The *Law of Intentionality* teaches us that growth doesn't just happen. You must design it, pursue it, and commit to it daily.

Your next chapter as a leader will not be written by accident. It will be written by your intentionality — by the choices you make each day to invest in clarity, resilience, influence, and growth.

The Walton Reminder

Sam Walton built Walmart by believing in people, experimenting relentlessly, and never losing sight of the customer. His principle still applies. Leaders who empower others and stay grounded in service will outlast those who cling to control.

Your slide decks or Jira dashboards will not measure your leadership brand — but by the trust you built, the systems you designed, and the people you grew.

A Final Charge

This isn't just a book to be read. It's a playbook to be lived. Choose one practice from each part — influence, resilience, scaling, future-proofing — and apply it this week. Small actions compound into big shifts.

Your career is your program. Your leadership is your brand. And your legacy will not be defined by the projects you delivered, but by the people, systems, and growth you enabled.

Go beyond delivery. Lead change. Grow with impact.

Appendix: TPM Leadership Framework

Purpose

The TPM Leadership Framework provides a repeatable model for how Technical Program Managers lead change, create clarity, and multiply impact without relying on formal authority.

1. Core Principles

- **Clarity over complexity** — Reduce noise, frame decisions, and simplify communication.
- **Influence over authority** — Earn trust through credibility, consistency, and contribution.
- **Resilience over reactivity** — Protect energy, perspective, and composure in high-pressure environments.
- **Growth over stagnation** — Continuously scale your systems, habits, and influence.

2. The TPM Leadership Flywheel

1. **Clarity** — Define the outcome, constraints, and decision owners.
2. **Cadence** — Establish the operating rhythm (meetings, forums, rituals).
3. **Visibility** — Create a single source of truth with the vital few metrics.
4. **Enablement** — Remove blockers, simplify dependencies, empower decision-making.
5. **Narrative** — Tell the story that aligns executives and teams.

As the flywheel spins, credibility compounds, and impact grows.

3. Practical Tools

- **Decision Map** — Who decides what, by when.
- **Risk Ledger** — Transparent ownership of risks and mitigations.
- **Program Charter Lite** — Crisp alignment on outcome, scope, and constraints.
- **Narrative Template** — Two slides: current state, risks/tradeoffs, recommendations.
- **Operating Cadence** — Predictable forums for alignment and decision-making.

4. Leadership Behaviors

- **Presence:** Stay calm, structured, and confident in the gray zone.
- **Framing:** Turn chaos into structured choices and tradeoffs.
- **Multiplication:** Build systems and artifacts that extend impact beyond yourself.
- **Character:** Lead with integrity, consistency, and service to others.

5. Self-Check Questions

- Am I creating clarity or adding noise?
- Do leaders trust me to frame decisions, not just report status?
- Have I protected my energy so I can lead with resilience?
- Am I scaling my impact through systems and artifacts?
- What story will be told about my leadership when I'm not in the room?

Use this framework as your leadership compass. Each time you begin a new program or step into a new role, return to these principles, flywheel steps, and tools. They are the difference between a TPM who delivers projects and a TPM who leads change.

Appendix: 30-Day Growth Plan for TPM Leaders

Purpose

This 30-day plan is designed to help Technical Program Managers apply the principles from *Beyond Delivery* in a practical, focused way. Each week has a theme, and each day offers a simple action. Use it as a daily compass — small, consistent steps build lasting growth.

Week 1: Clarity and Framing

Theme: Practice creating clarity in every interaction.

- **Day 1:** Write down the top three outcomes for your current program. Share them with your team.
- **Day 2:** Map the decision owners for one active initiative.
- **Day 3:** Replace one status update with a decision-framing narrative.
- **Day 4:** Ask a stakeholder, *"What's unclear for you right now?"* — then close the gap.
- **Day 5:** Simplify a slide or report into one-page clarity.
- **Day 6:** Reflect: Where did I reduce noise today? Where did I add it?
- **Day 7:** Journal: What does "creating clarity" look like for me as a leader?

Week 2: Influence and Trust

Theme: Build credibility through competence, consistency, and contribution.

- **Day 8:** Identify one area where you can deliver early to build trust.
- **Day 9:** Shadow or interview a peer to understand their perspective better.
- **Day 10:** Practice "listen first" in every meeting today.
- **Day 11:** Share an insight (not just status) in an update to leadership.
- **Day 12:** Thank someone publicly for their contribution.
- **Day 13:** Reflect: Did I keep my commitments this week, big and small?
- **Day 14:** Journal: What three words do I want people to associate with me?

Week 3: Resilience and Energy

Theme: Protect your routines and sustain your leadership presence.

- **Day 15:** Block 30 minutes for reflection or journaling.
- **Day 16:** Walk or exercise before your first meeting.
- **Day 17:** Prioritize sleep tonight — schedule it like a meeting.
- **Day 18:** After a tough moment, pause and ask: *"What's the real risk here?"*
- **Day 19:** Delegate one task you don't need to carry alone.
- **Day 20:** Reflect: How did my energy impact my presence this week?
- **Day 21:** Journal: Where am I overextending? Where do I need renewal?

Week 4: Scaling and Future-Proofing

Theme: Multiply your impact and prepare for the future.

- **Day 22:** Identify one recurring issue and design a simple system to fix it.
- **Day 23:** Automate or delegate a repetitive task.
- **Day 24:** Draft a one-page narrative connecting delivery to business outcomes.
- **Day 25:** Experiment with an AI tool to synthesize notes or risks.
- **Day 26:** Mentor a peer or junior TPM — share one tool or framework.
- **Day 27:** Reflect: What artifacts carry my leadership brand beyond the room?
- **Day 28:** Journal: What legacy do I want my leadership to leave here?
- **Day 29:** Write down three beyond-corporate experiments you could try.
- **Day 30:** Celebrate — capture your key wins from this month and set 3 growth goals for the next 90 days.

How to Use This Plan

- Pick one small action each day — don't skip ahead.
- Reflect weekly on what changed in your clarity, influence, energy, or scaling.
- Repeat the cycle quarterly — each pass deepens your leadership habits.

Appendix: Reflection Prompts & Journaling Exercises

Purpose

Leadership growth begins with self-awareness. These reflection prompts and journaling exercises are designed to help TPMs move from theory to practice — turning daily experiences into leadership lessons. Use them weekly, monthly, or after major program milestones.

Section 1: Clarity & Influence

- When was the last time I created clarity in a chaotic situation? How did others respond?
- What is one decision I can frame more clearly this week?
- Do my updates focus on activity or outcomes? How can I reframe them?
- Who trusts me most right now? Why? Who do I need to rebuild trust with?
- Journaling Exercise: Write a 1-page reflection on how I want others to describe my leadership brand in 12 months.

Section 2: Resilience & Energy

- Where am I currently overextended?
- What routines protect my energy? Which ones have I neglected?
- How do I typically react under pressure? How do I want to react?

- Am I building time for reflection and renewal, or am I constantly firefighting?
- Journaling Exercise: Track my energy across 3 days. What activities drain me? What activities restore me?

Section 3: Scaling Impact

- What system or cadence could I design once that would save hours every week?
- Do my artifacts (dashboards, narratives, charters) reflect my leadership brand?
- Am I a bottleneck, or am I building systems that work when I'm not in the room?
- Where can I leverage AI or data to multiply my reach?
- Journaling Exercise: Write a 1-page description of what "force multiplier" leadership looks like in my role.

Section 4: Future-Proofing

- What skills do I need to develop in the next 12 months to stay relevant?
- Where do I see my career 5 years from now? 10 years from now?
- What fears are holding me back from exploring opportunities beyond corporate?
- Am I investing in mentoring others and building leaders who can scale impact?
- Journaling Exercise: Draft a "Beyond Corporate Experiment" — one way I could apply TPM skills outside my company.

How to Use This Appendix

1. **Pick one question per week** — use it as your journal prompt.
2. **Block 15–20 minutes** of uninterrupted time. Write freely, don't edit.
3. **Review monthly** — look back for patterns in your responses.
4. **Share selectively** — discuss insights with a mentor, coach, or trusted peer to accelerate growth.

Remember: Growth isn't automatic. It's intentional. Reflection turns experience into insight, and insight into action.

Appendix: Resources and Communities

Purpose

No TPM leads alone. Growth accelerates when you tap into resources, learn from others, and connect with communities that share your challenges and goals. This appendix provides a curated list of books, courses, tools, and networks to help you continue your journey **beyond delivery**.

Books & Thought Leadership

- *The 15 Invaluable Laws of Growth* — John C. Maxwell (personal growth principles that apply directly to leadership).
- *Developing the Leader Within You 2.0* — John C. Maxwell (shaping influence and character).
- *Made in America* — Sam Walton (practical leadership and culture-building lessons).
- *The Lean Startup* — Eric Ries (entrepreneurial systems thinking).
- *Measure What Matters* — John Doerr (OKRs and outcome-focused leadership).

Courses & Certifications

- **Maxwell Leadership Certified Team (MLCT):** Training on leadership, communication, and coaching. You can contact me for a referral.
- **Project Management Institute (PMI):** PMP and PgMP certifications for program leadership.

- **Scrum Alliance:** CSM or SAFe Program Consultant for agile scaling frameworks.
- **Coursera TPM Courses:** Business strategy, AI for leaders, and systems thinking.

Communities & Networks

- LinkedIn Groups:
    - Technical Program Managers Network
    - Leadership in Tech
- Professional Associations:
    - *PMI Chapters* (local networking and learning events).
    - *Women in Product / Women in TPM* (for mentorship and career growth).
- Conferences:
    - Maxwell Leadership International Conference (IMC)
    - Agile Alliance Conference
    - O'Reilly AI & Systems Conference

Tools & Templates

- **AI Productivity Tools:** Notion AI, ChatGPT, Jasper — for drafting, synthesis, and decision framing.
- **Program Management Tools:** Jira, Confluence, Trello, Asana — for execution and tracking.
- **Visualization Tools:** Miro, Figma, Lucidchart — for mapping dependencies and storytelling.
- **Analytics & Dashboards:** Power BI, Tableau, Google Data Studio — for visibility and metrics clarity.

Podcasts & Media

- *Maxwell Leadership Podcast* — practical leadership principles.
- *Masters of Scale* — Reid Hoffman on building systems and organizations.
- *HBR IdeaCast* — leadership and organizational strategy.
- *The Modern Manager* — tactical advice for leading with clarity and empathy.

How to Use This Appendix

1. **Pick one book and one community** to engage within the next 30 days.
2. **Leverage tools intentionally** — focus on clarity, not complexity.
3. **Invest in people** — the best resources are the mentors and peers who walk with you.

Leadership is a journey best taken in community. Surround yourself with resources, mentors, and peers who sharpen your thinking, strengthen your resilience, and multiply your impact.

References & Further Reading

The following works influenced the principles in this book and are recommended for further study. Each offers valuable insights into leadership, growth, and organizational impact.

Maxwell, John C. *The 21 Irrefutable Laws of Leadership: Follow Them and People Will Follow You*. Thomas Nelson, 2007.

Maxwell, John C. *Failing Forward: Turning Mistakes into Stepping Stones for Success*. Thomas Nelson, 2000.

Maxwell, John C. *The 15 Invaluable Laws of Growth: Live Them and Reach Your Potential*. Center Street, 2012.

Walton, Sam, and John Huey. *Made in America: My Story*. Doubleday, 1992.

Appendix E: Field Applications & Templates

Program Charter Lite

- Outcome:
- Constraints (Top 3):
- Key Decisions & Owners:
- Success Metrics:

Decision Map

- Decision:
- Owner (Decider):
- Contributors:
- Input Artifacts:
- Options & Tradeoffs:
- Deadline:
- Status:
- Outcome & Rationale:

Risk Ledger

- Risk:
- Impact:
- Likelihood:
- Owner:
- Mitigation Plan:
- Status:

Make this list as table columns in Confluence or Excel. Use each row to drive conversation, clarity, and make decisions for mitigation.

Weekly Decision Review Template

- Date:
- Key Risks:
- Options:
- Tradeoffs:
- Recommendation:
- Decision Taken:

About the Author

Johnathan Stephen Sexton has over two decades of experience leading large-scale digital transformations at Walmart Global Tech. He has driven initiatives spanning pharmacy delivery, replenishment systems, and enterprise data platforms, earning recognition for creating clarity, building influence, and enabling execution at scale.

A Maxwell Leadership Certified coach and speaker, he blends real-world corporate experience with timeless leadership principles to help others grow in clarity, resilience, and influence. Johnathan is also a certified Project Management Professional (PMP) and NASM Certified Personal Trainer (CPT), equipping him to guide leaders in both professional and personal growth.

Beyond his corporate career, Johnathan is the founder of Flesh Tamers Fitness, where he integrates faith, fitness, and leadership coaching to equip leaders for holistic growth.

A marathoner, writer, and mentor, Johnathan helps leaders strengthen not only their organizations but also their lives — mentally, physically, and spiritually. His mission is to empower leaders to move beyond delivery into lasting impact.

Table of Contents